Atlas of Microbiology

A. Fuad Khan

Learning Solutions

New York   Boston   San Francisco
London   Toronto   Sydney   Tokyo   Singapore   Madrid
Mexico City   Munich   Paris   Cape Town   Hong Kong   Montreal

Cover Art: A. Fuad Khan

Printed in the United States of America

3 4 5 6 7 8 9 10 0BRV 14 13 12

2009140561

CY

www.pearsonhighered.com

ISBN 10: 0-558-44063-0
ISBN 13: 978-0-558-44063-3

Dedication

To my parents and to my brother Shahzad.

Contents

1. Stained slides

R. nigricans, T. gambiense, and Human blood. 1

2. Aseptic Technique

Culture of S. marcescens on nutrient agar slant, nutrient agar deep tube, and in nutrient broth . . . 2,3

3. Simple Stain

B. cereus, S. epidermidis, and E. coli 4

4. Gram Stain

B. cereus, S. epidermidis, and E. coli 5

5. Negative Stain

B. cereus and Micrococcus luteus. 6

6. Acid-Fast Stain (Ziehl-Neelson Method)

Mycobacterium smegmatis and Staphylococcus epidermidis 7

7. Endospore Stain (Schaffer-Fullton Method)

Bacillus cereus and Clostridium botulinum 8

8. Capsule Stain

Klebsiella pneumoniae. 9

9. Cultural Characteristics of Selected Bacteria

B. cereus, E. coli, M. luteus, P. aeruginosa, S. aureus 10-13

10. Enumeration of Viable Bacteria

Serial Dilution Method. 14-16

11. Selective and Differential Media

Mannitol Salt Agar 17
MacConkey Agar. 18
Eosin-Methylene Blue Agar 19
Blood Agar 20
Phenylethyl Alcohol Agar 21

12. Cultivating Anaerobic Microbes

GasPak Jar Method 22
Reducing Medium Method 23

13. Effect of Ultraviolet Radiation on Selected Bacteria

Bacillus cereus, E. coli, and S. aureus 24, 25

14. Fungi: Yeast and Mold

R. nigricans, Aspergillus niger, Penicillium notatum 26-30

15. Protozoa

Free living 31
Parasitic 32-34

16. Parasitic Helminthes

Tapeworm (Taenia pisiformis) 35-37
Liver fluke (Fasciola hepatica) 38
Blood fluke (Schistosoma japonicum) 39, 40
Pinworm (Enterobius vermicularis) 41, 42
Hookworm (Necator americanus) 43, 44
Ascaris lumbricoides 45

17. Effect of Antibiotics on Bacteria

S. aureus, E. coli, P. aeruginosa, and M. smegmatis 46-53

18. Effect of Antiseptics and Disinfectants on Bacteria

S. aureus, E. coli, P. aeruginosa, and M. smegmatis 54-56

19. Bacterial Conjugation

E. coli (Hfr and F-) 57

20. IMViC Test

Indol test 58-61

Methyl Red test 59

Voges-Prauskauer test 59

Citrate test 60

21. Carbohydrate Fermentation

Escherichia coli (Lactose, Galactose, and Sucrose) 62

Pseudomonas aeruginosa (Lactose, Galactose, and Sucrose) 63

Staphylococcus aureus (Lactose, Galactose, and Sucrose) 64

22. Triple Sugar Iron Test

E. coli, P. vulgaris, P. aeruginosa, S. typhimurium, and Shigella dysenteriae 65, 66

23. Extra-cellular Enzymes

Gelatin Hydrolysis 67

Starch Hydrolysis 68

Casein Hydrolysis 69, 70

24. Nitrate Reduction Test

Alcaligenes faecalis, Escherichia coli, and Pseudomonas aeruginosa 71, 72

25. Normal Flora of the Skin

Mannitol Salt Agar plates 73

Blood Agar plates 74

Sabouraud's agar plates 75

26. Catalase Test

Staphylococcus aureus, Micrococcus luteus, and Streptococcus lactis 76

27. Qualitative Analysis of Milk

Methylene Blue Reduction test 77

Index 79-81

Preface

In life sciences like Zoology, Botany and Human Anatomy, students can see many details with their naked eye. Microbiology is unique among the life sciences because its details, microorganisms, cannot be seen with naked eye. This Atlas of Microbiology is intended for students enrolled in introductory courses of microbiology. One of the challenges teaching introductory microbiology courses is the difficulty many students have making accurate observations on a consistent basis. This is mainly due to failure to use aseptic techniques. This atlas of microbiology allows students to compare their observations and results with the accurate observations and results included in this atlas.

This atlas is a direct result of my sincere interest in teaching and is based on many years of experience teaching undergraduate microbiology. The idea for this atlas was conceived many years ago and has finally gone from idea to fruition.

Acknowledgements

I wish to thank my highly dedicated lab staff Wendy and Tracie for their complete and constant support. I also want to express my deep appreciation to my wife, Hina Khan, my son, Nabeel Khan, and my daughter Natasha Khan, for their help, patience and encouragement through the many months of hard work.

Finally, I wish to express my appreciation to my colleagues Teresa Diehl and David Cook for their technical assistance with photography and photo editing.

Stained Slides

R. Nigricans (100X)

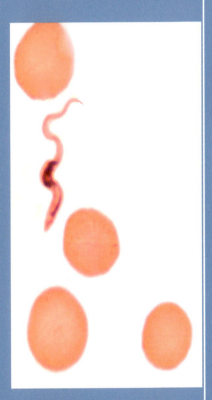

Human blood (1000X)

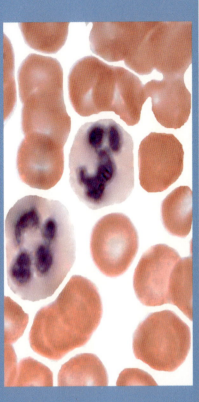

T. gambiense (1000X)

10 X 10   10 X 100   10 X 100

Magnification = (magnification of ocular lens **X** magnification of objective lens)

X = magnification

Aseptic Technique

Serretia marcescens

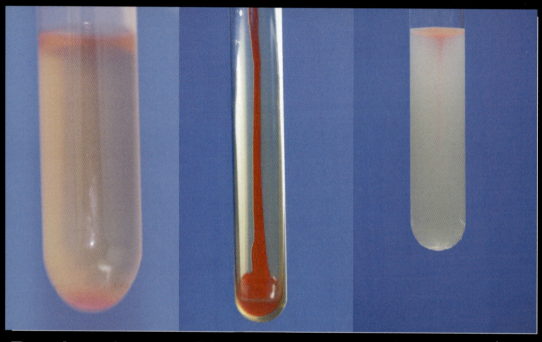

Broth culture **Slant culture** **Deep tube**

(after 48 hours incubation at 25°C)

Aseptic Technique
Isolation of Pure culture:
Streak Plate Method

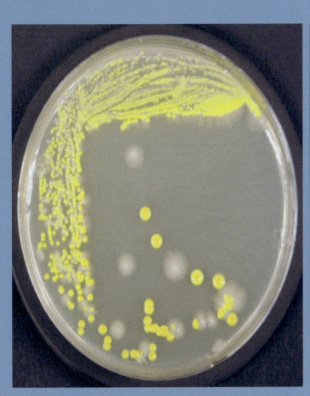
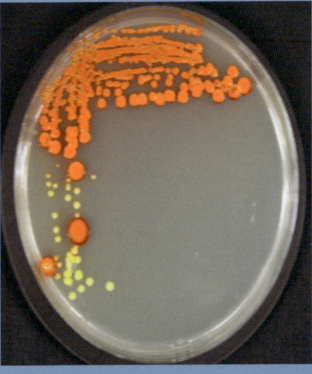

E. coli and M. luteus   **E. Coli and S. marcescens**
(after 48 hours incubation at 25°C)

M. luteus is bright yellow
S. marcescens is redish orange and
E. coli is grayish white

Simple Stain

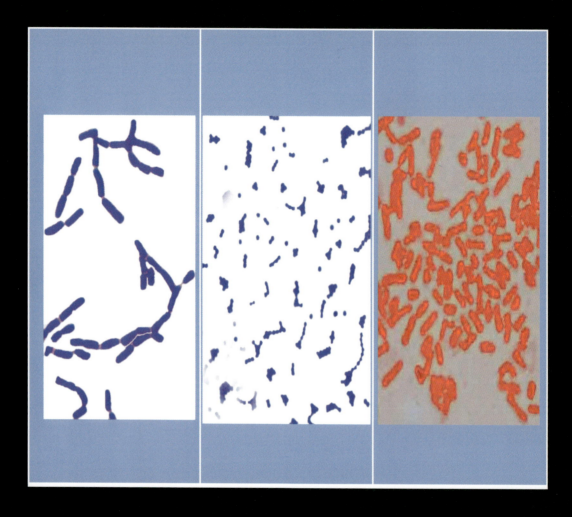

B. cereus   S. epidermidis   E. coli

B. cereus: stained with Methylene Blue stain. It is bacillus, rod shape, and grows in chains

S. epidermidis: stained with Crystal violet stain. It is coccus, round/spherical shape and grows in bunches.

E. coli: stained with Safranin. It is short bacillus and brows as diplobacilli.

Gram Stain

Gram positive	Gram positive & Gram Negative	Gram Negative
B. cereus	S. aureus and E. coli	E. coli

- Young Gram- positive bacteria , because of high peptidoglycan and low lipid in their cell wall, retain primary stain and appear purple
- Old cultures of Gram-positive bacteria and over decolorization may cause them to appear red/pink instead of purple. This is called Gram variability.
- Gram-negative bacteria, because of low peptidoglycan and high lipid concentration in their cell wall, lose the primary color but hold the counter stain, safranin, and appear red/pink.

Negative Stain

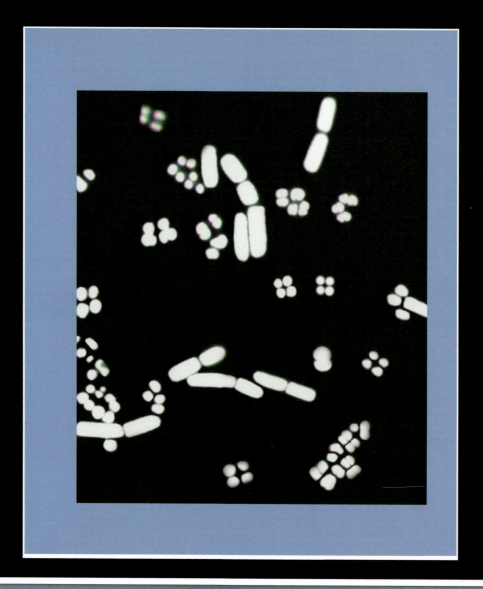

Both the stain, Nigrosin, and the bacteria are negatively charged. Therefore the bacteria remained colorless and the background is black.

M. luteus: coccus shape, grows in tetrads
B. cereus: bacillus shape, grows in chains

Acid-fast Stain

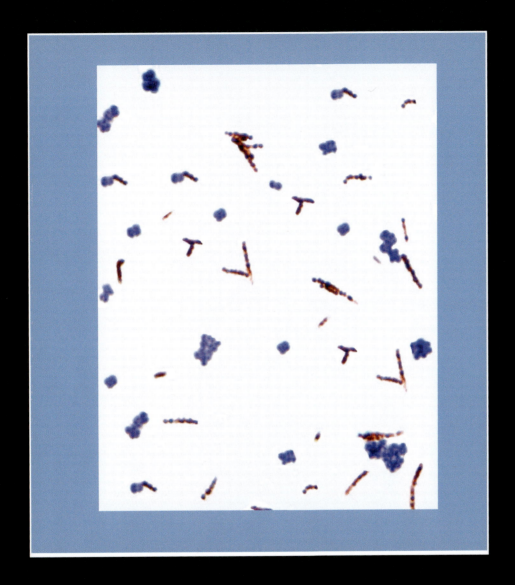

Mycobacterium smegmatis: acid-fast bacterium, magenta color, pleomorphic bacilli

Micrococcus luteus: non-acidfast bacterium, grows in tetrads (groups of four cocci).

Endospore Stain
Schaffer-Fullton Method

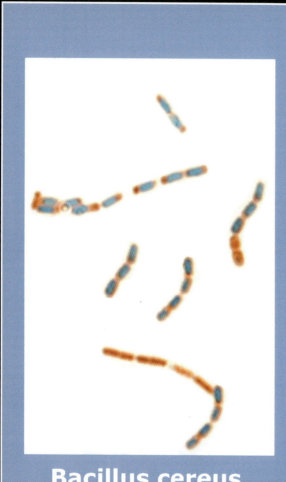

Bacillus cereus

Clostridium botulinum

B. cereus
- Endospores retain the primary stain malachite green during steaming and appear green.
- Vegetative cells hold the secondary stain and appear red.

Clostridium botulinum
This is a simple stain of endospores. The endospores remain clear. The vegetative cell is purple.

Capsule Stain

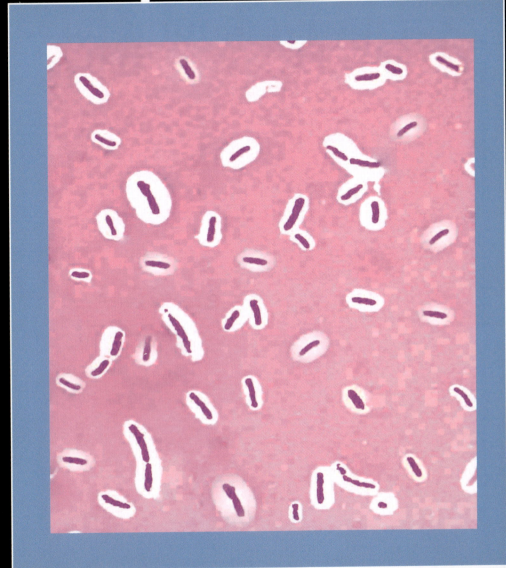

Klebsiella pneumoniae

- Purple bacilli are bacteria
- Clear halo around purple bacilli are capsules
- Capsules are non-ionic, made of polysaccharides, proteins, or glycoprotein and makes bacteria virulent.

Cultural Characteristics
Escherichia coli

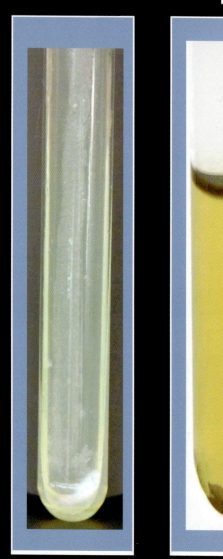

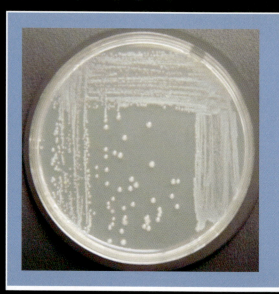

c

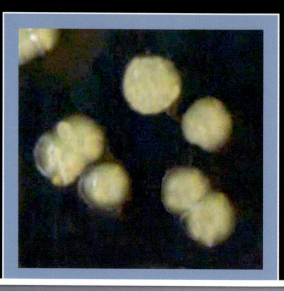

d

a

b

a= growth on nutrient agar slant
b= growth in nutrient broth
c= growth on nutrient agar plate
d= isolated colonies on nutrient
    agar plate

Cultural Characteristics
Bacillus cereus

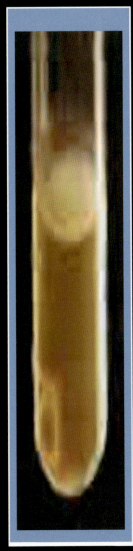

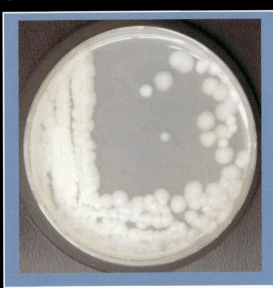

c

d

a

b

a= growth on nutrient agar slant
b= growth in nutrient broth
c= growth on nutrient agar plate
d= isolated colonies on nutrient
   agar plate

Cultural Characteristics
Micrococcus luteus

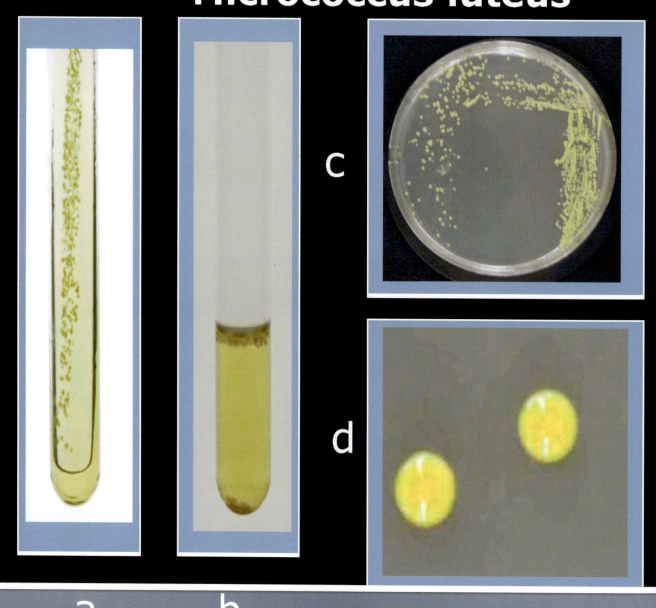

c

d

a

b

a= growth on nutrient agar slant
b= growth in nutrient broth
c= growth on nutrient agar plate
d= isolated colonies on nutrient
    agar plate

Cultural Characteristics
Pseudomona aeruginosa

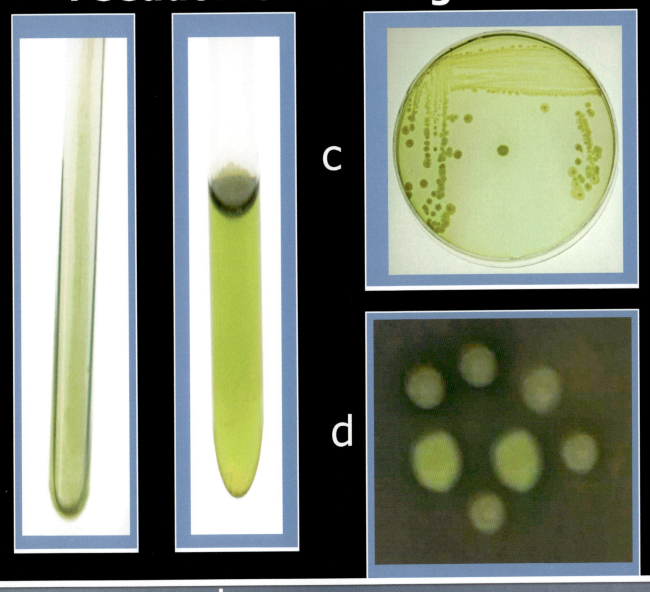

a

b

c

d

a= growth on nutrient agar slant
b= growth in nutrient broth
c= growth on nutrient agar plate
d= isolated colonies on nutrient
   agar plate

Enumeration of Bacteria: Serial Dilution Method

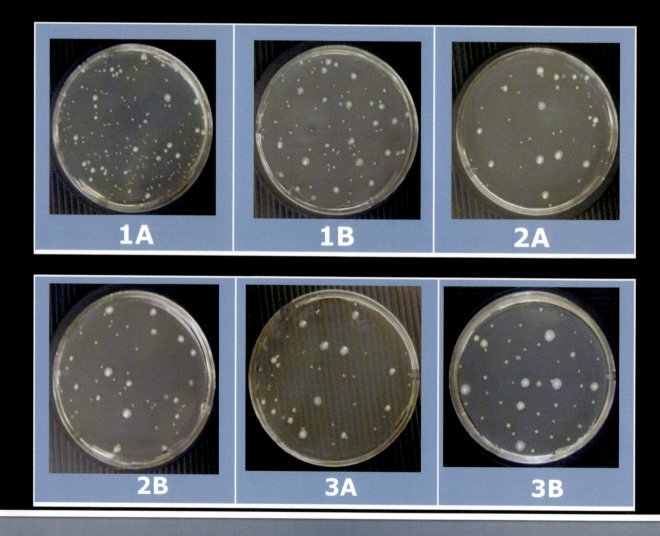

1A= Too Numerous to Count (TNTC)

Enumeration of Bacteria: Serial Dilution Method

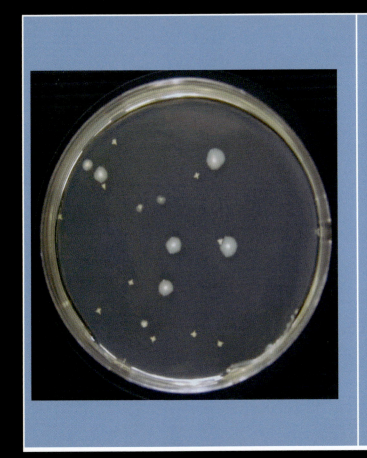

Too Few To Count
(TFTC)

Too Numerous To Count
(TNTC)

Enumeration of Bacteria:

Serial Dilution Method

Qubec Colony Counter

Selective and Differential Media
Mannitol Salt Agar (MSA)

S. aureus S. epidermidis

S. aureus: is a facultative halophilic bacterium. It ferments mannitol and produces acid which turns the color of the mannitol salt agar medium from pink to yellow.
S. epidermidis: is also a facultative halophilic bacterium. But it cannot ferment mannitol, thus the color of the medium remains pink.

Selective and Differential Media
MacConkey Agar

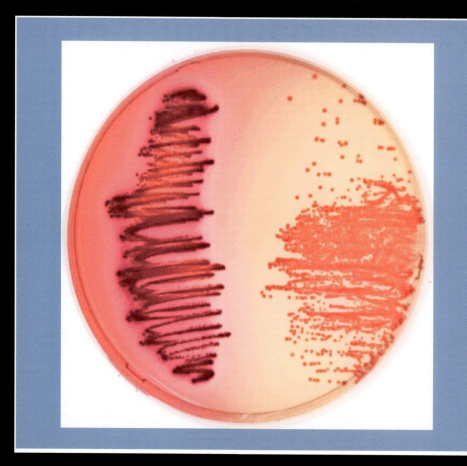

E. Coli   E. aerogenes

E. coli: Is Gram-negative and ferments lactose and produces large quantities of acid that is indicated by the pink halo around the pink growth of the bacterium

E. aerogenes: is a Gram-negative and ferments lactose but produces small quantities of acid from lactose . The growth remains pink but there is no halo around the growth.
Gram positive bacteria do not grow on this medium.

Selective and Differential Media
Eosin-Methylene Blue Agar (EMB)

E. Coli   S. typhimurium

E. coli: Gram-negative and ferments lactose It produces a green-metallic sheen on this medium

S. typhimurium: Gram-negative and does not ferments lactose. The growth remains colorless/graish.

Gram-positive bacteria grow poorly on this medium

Selective and Differential Media

Blood Agar

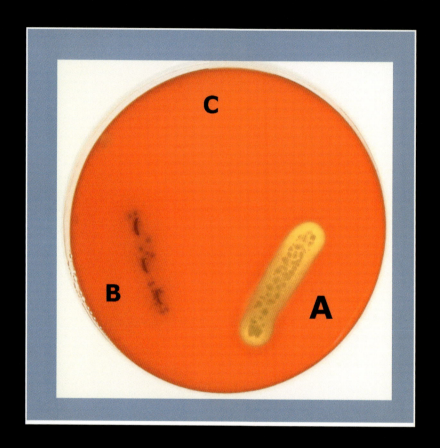

A= S. Pyogenes: Beta-hemolysis. Clear zone around the growth indicates complete digestion of hemoglobin

B= E. faecalis: Alpha-hemolysis. Greenish zone around the growth indicates incomplete/partial digestion of hemoglobin

C= S. mitis: Gamma- hemolysis. No zone around the growth indicates no digestion of hemoglobin.

Selective and Differential Media
Phenylethyl Alcohol Medium (PEA)

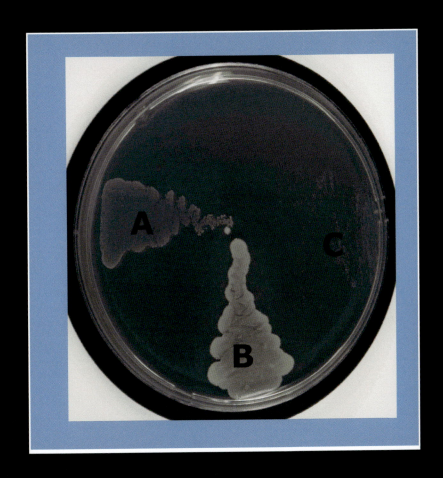

A= E. faecalis: Gram-positive. Grows well on this medium

B= S. aureus: Gram-positive. Grows well on this medium

C= E. coli: Gram-negative. Grows poorly on this medium

Cultivating Anaerobic Microbes

Sodium Thioglycollate Broth

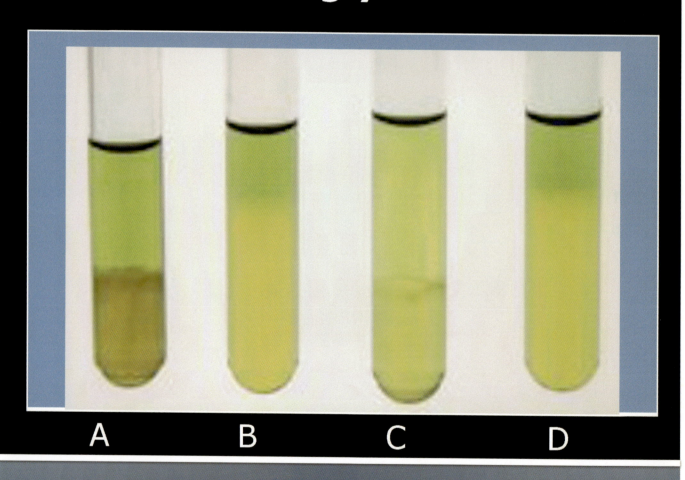

A   B   C   D

A = Clostridium sporogenes (growth present)
B = Control
C = Escherichia coli (growth present)
D = Micrococcus luteus (no growth)

Cultivating Anaerobic Microbes

Plate incubated in GasPak Jar

Plate incubated Aerobically

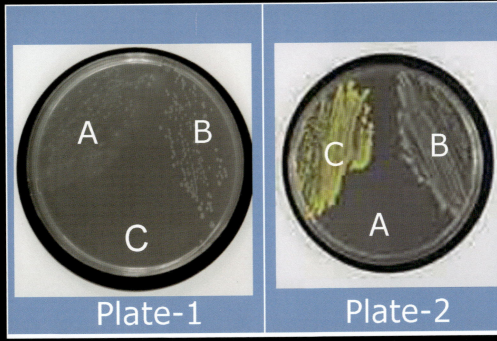

Plate-1

Plate-2

A = Clostridium sporogenes
B = Escherichia coli
C = Micrococcus luteus

Clostridium sporogenes: can be classified as an anaerobe as it only grew on plate-1 (in GasPak Jar).
Escherichia coli: can be classified as facultative anaerobe as it grew on both plates but grew better on plate-2 (incubated aerobically)
Micrococcus luteus: can be classified as aerobe as it only grew on plate-2 (incubated aerobically).

Effect of Ultraviolet Radiation on Bacteria

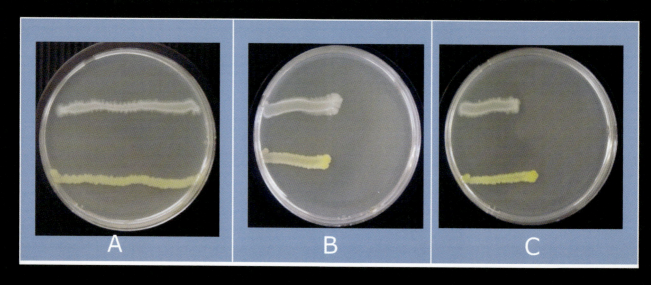

E. Coli (grayish-white growth, Gram-negative, non-endospore producer) and
S. aureus (Gram-positive, non-endospore producer, golden-yellow growth)
A = unexposed control
B = 30 seconds (left side unexposed and right side exposed)
C = 1 minute (left side unexposed and right side exposed)
D = 3 minutes (left side unexposed and right side exposed)
E = 3 minutes, entire plate was exposed with lid on

Effect of Ultraviolet Radiation on Bacteria

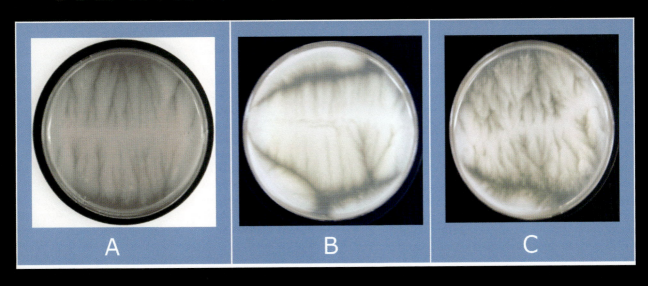

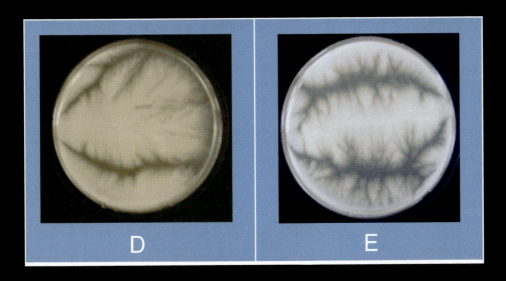

Bacillus cereus (Gram-positive, endospore producer)

A = unexposed control
B = 30 seconds (left side unexposed and right side exposed)
C = 1 minute (left side unexposed and right side exposed)
D = 3 minutes (left side unexposed and right side exposed)
E = 3 minutes, entire plate was exposed with lid on

Fungi: Yeast and Mold

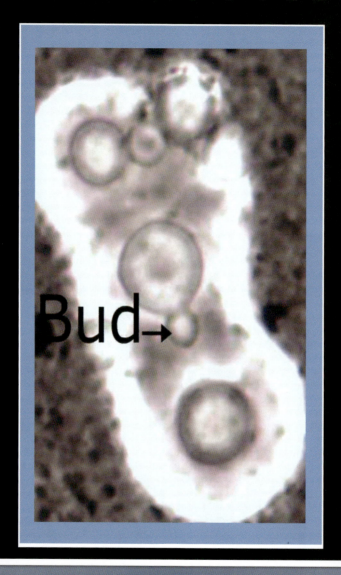

Cryptococcus neoformans (yeast)

Fungi: Yeast and Mold

Conidiospores/conidia (asexual spores)

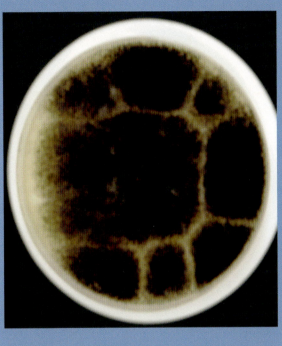

Aspergillus niger (1000X) (Mold)

Aspergillus niger on Sabouraud's agar

Fungi:
Yeast and Mold

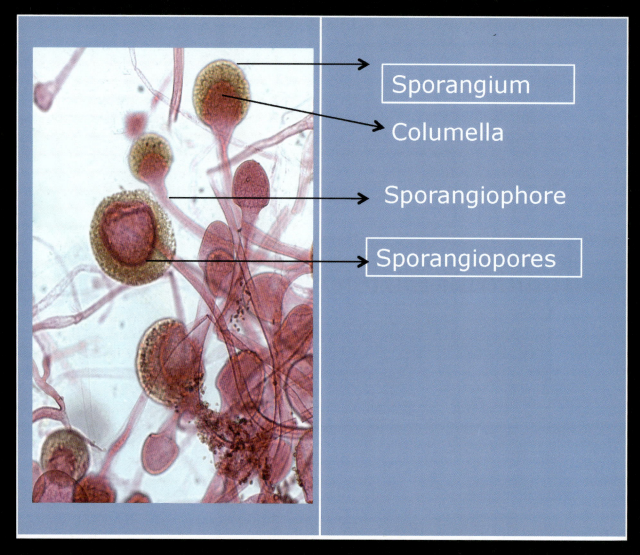

Sporangium

Columella

Sporangiophore

Sporangiopores

Rhizopus nigricans (100X)
(Mold)

Fungi:
Yeast and Mold

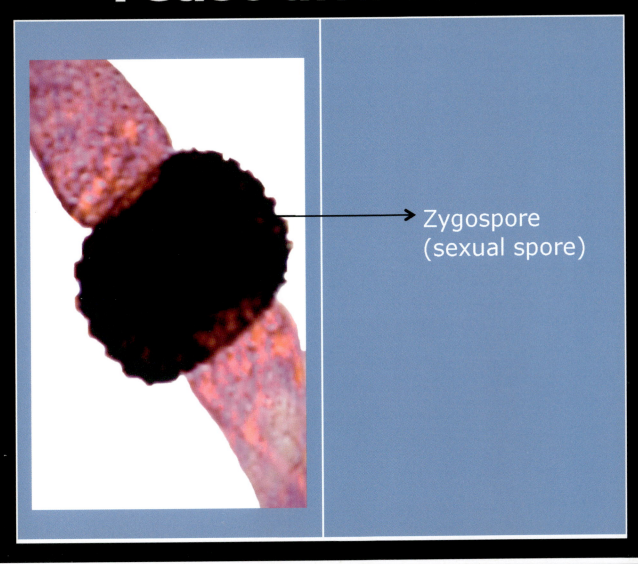

Zygospore
(sexual spore)

Rhizopus nigricans (100X)
(Mold)

Fungi: Yeast and Mold

Conidiospores/conidia (asexual spores)

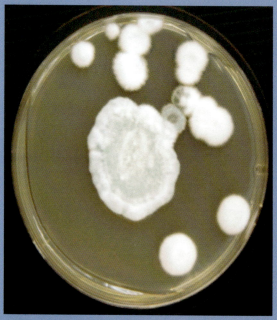

Penicillium notatum (1000X) (mold)

Penicillium notatum on Sabouraud's agar

Free Living Protozoa

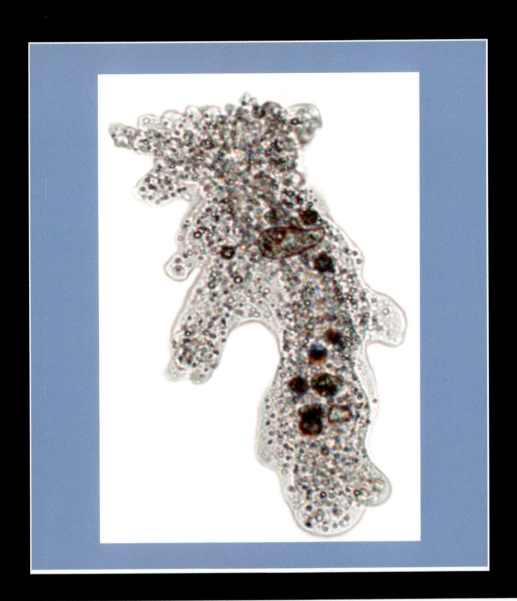

Amoeba proteus

Parasitic Protozoa

Class: Sarcodina: Entamoeba histolytica

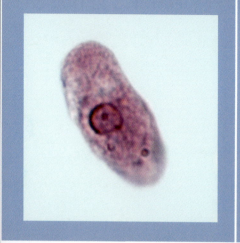 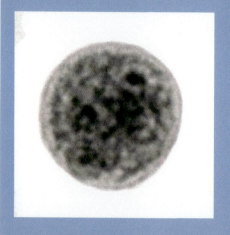

Trophozoite   Cyst

Class: Mastigophora: Giardia lambliam

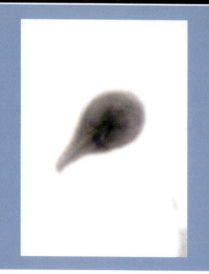 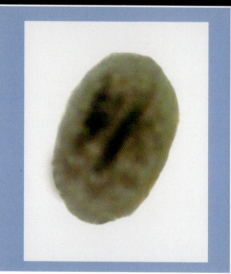

Trophozoite   Cyst

Parasitic Protozoa
Plasmodium vivax

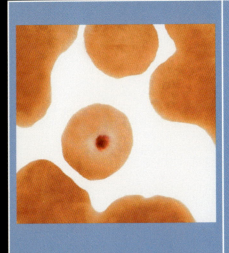 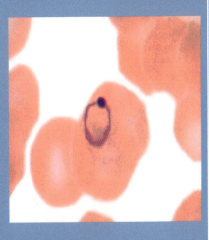 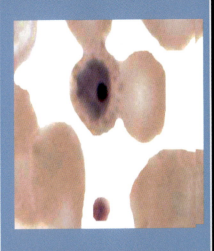

Merozoite   Signet Ring   Trophozoite

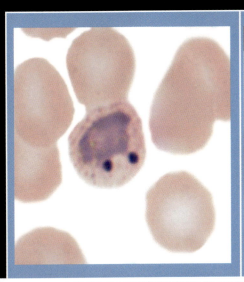 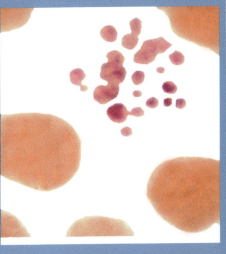

Early Schizont Late Schizont

Parasitic Protozoa
Balantidium coli

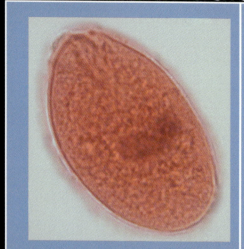
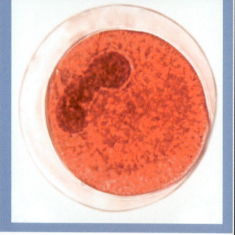

Trophozoite   Cyst

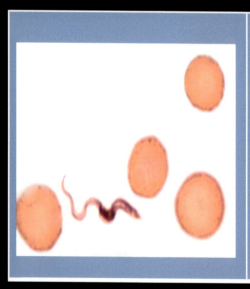
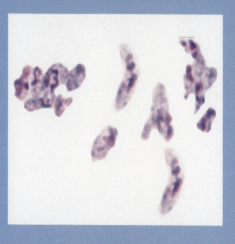

Trypanosoma
gambiense

Toxoplasma
Gondii
(trophozoite

34

Parasitic Helminthes

Phylum: Platyhelminthes
Class: Cestoda

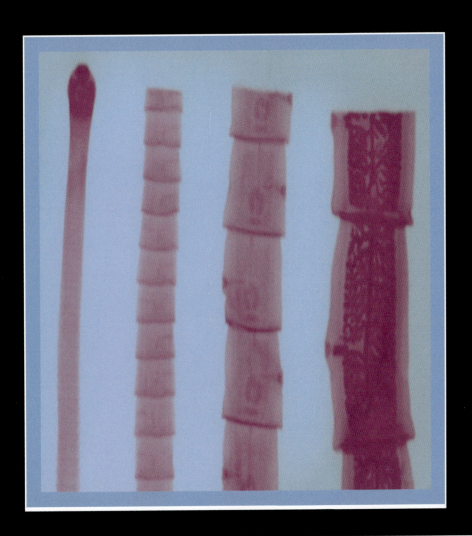

Developmental stages of Taenia pisiformis (dog tapeworm)

Parasitic Helminthes
Phylum: Platyhelminthes
Class: Cestoda

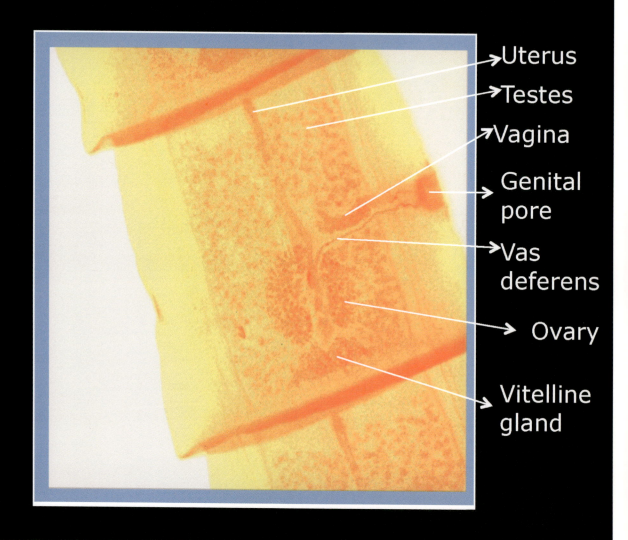

Uterus
Testes
Vagina
Genital pore
Vas deferens
Ovary
Vitelline gland

Mature Proglottid of dog tapeworm (Taenia pisiformis) showing male and female reproductive gonads.

Parasitic Helminthes

Phylum: Platyhelminthes
Class: Cestoda

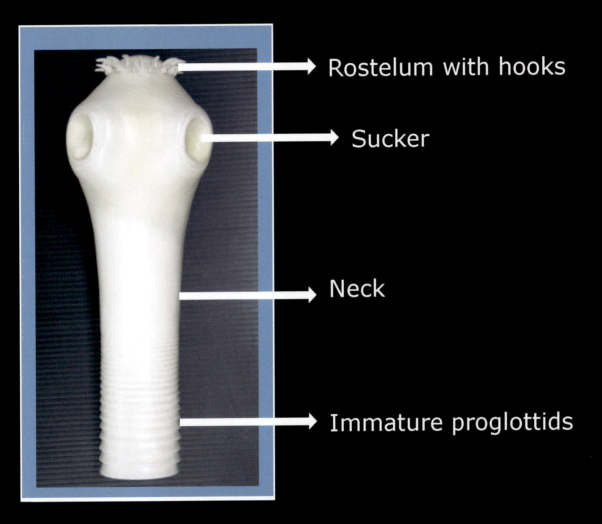

Rostelum with hooks

Sucker

Neck

Immature proglottids

Scolex model of tapeworm

Parasitic Helminthes
Phylum: Platyhelminthes
Class: Trematoda

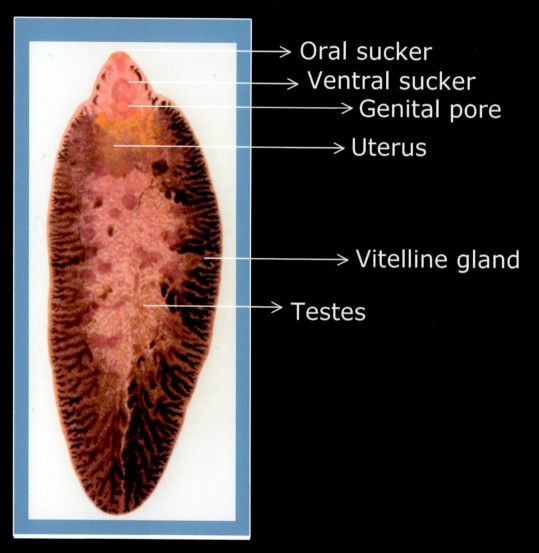

→ Oral sucker
→ Ventral sucker
→ Genital pore
→ Uterus
→ Vitelline gland
→ Testes

Fasciola hepatica (liver fluke)

Parasitic Helminthes
Phylum: Platyhelminthes
Class: Trematoda

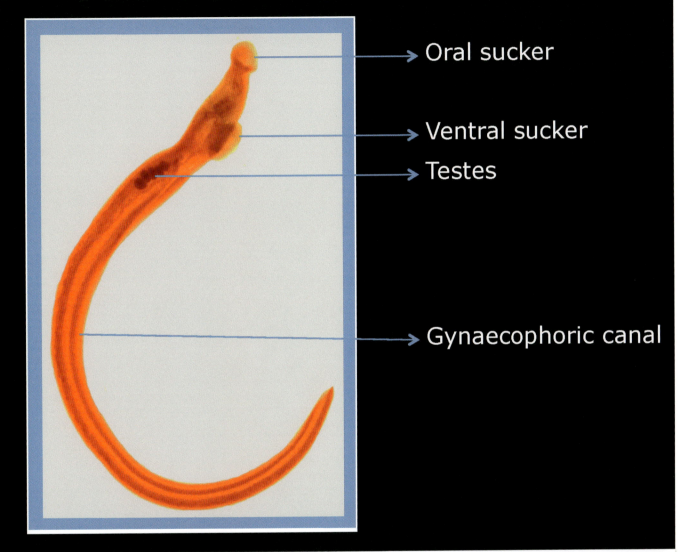

Oral sucker

Ventral sucker

Testes

Gynaecophoric canal

Schistosoma japonicum (blood fluke)
Male

Parasitic Helminthes
Phylum: Platyhelminthes
Class: Trematoda

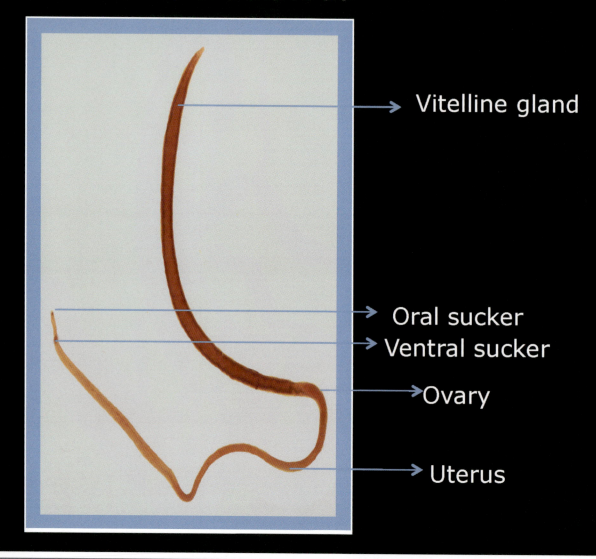

Vitelline gland

Oral sucker
Ventral sucker
Ovary
Uterus

Schistosoma japonicum (blood fluke)
Female

Parasitic Helminthes
Phylum: Nemathelminthes
Class: Nematoda

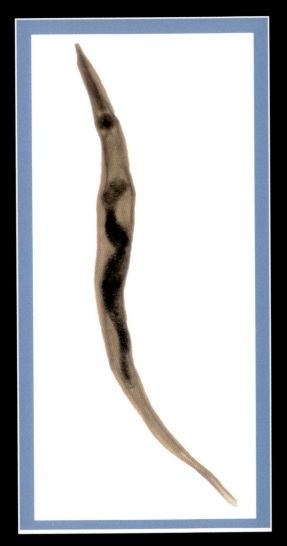

**Enterobius vermicularis (pinworm)
Male**

Parasitic Helminthes
Phylum: Nemathelminthes
Class: Nematoda

**Enterobius vermicularis (pinworm)
Female**

Parasitic Helminthes

Phylum: Nemathelminthes
Class: Nematoda

**Necator americanus (hookworm)
Male**

Parasitic Helminthes

Phylum: Nemathelminthes
Class: Nematoda

**Necator americanus (hookworm)
Female**

Parasitic Helminthes

Phylum: Nemathelminthes
Class: Nematoda

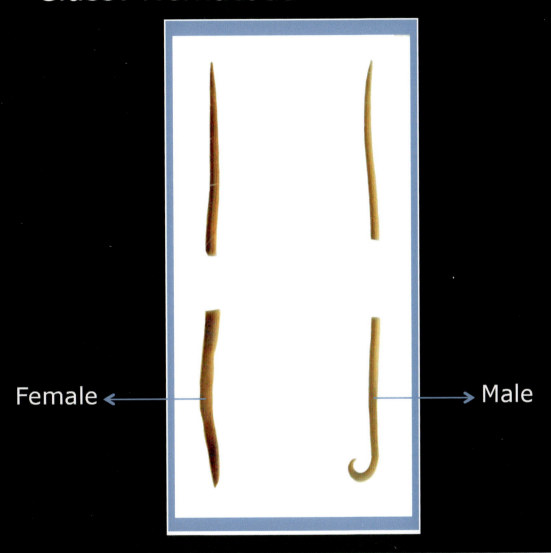

Female

Male

Ascaris lumbricoides

Chemotherapeutic Agents
Antibiotics

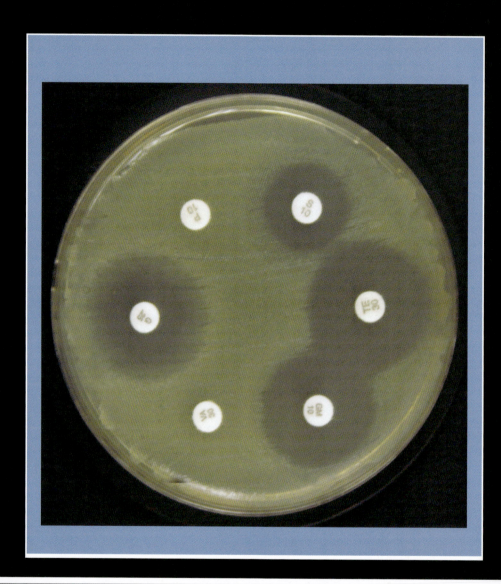

Staphylococcus aureus
(Gram-positive)

P=	Penicillin
S=	Streptomycin
Te=	Tetracycline
GM=	Gentamycin
VA=	Vancomycin
G=	Sulfisoxazole

Chemotherapeutic Agents
Antibiotics

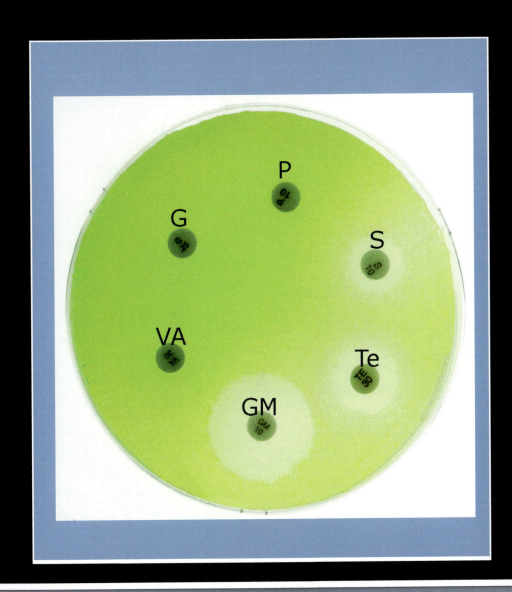

Pseudomonas aeruginosa
(Gram-negative)

P=	Penicillin
S=	Streptomycin
Te=	Tetracycline
GM=	Gentamycin
VA=	Vancomycin
G=	Sulfisoxazole

Chemotherapeutic Agents Antibiotics

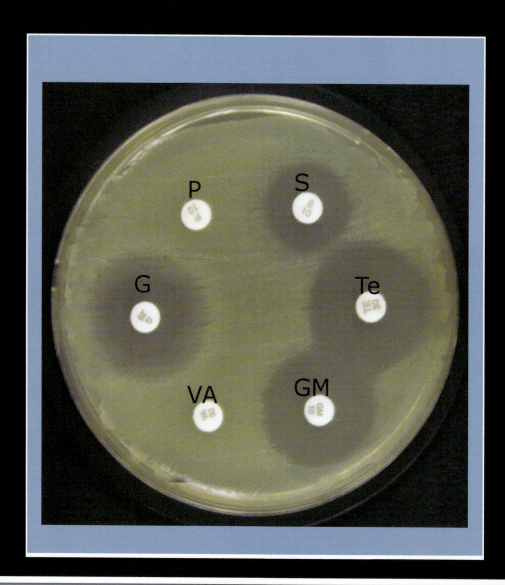

Escherichia coli
(Gram-negative)

P=	Penicillin
S=	Streptomycin
Te=	Tetracycline
GM=	Gentamycin
VA=	Vancomycin
G=	Sulfisoxazole

Chemotherapeutic Agents Antibiotics

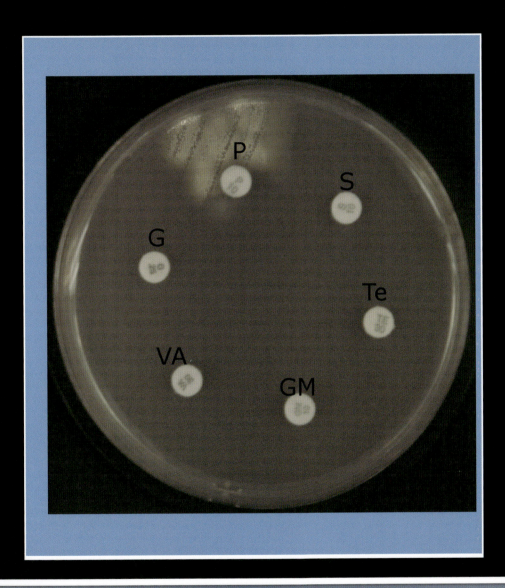

Mycobacterium smegmatis
(Acid-fast)

P=   Penicillin
S=   Streptomycin
Te=   Tetracycline
GM=   Gentamycin
VA=   Vancomycin
G=   Sulfisoxazole

Chemotherapeutic Agents
Additive Effect

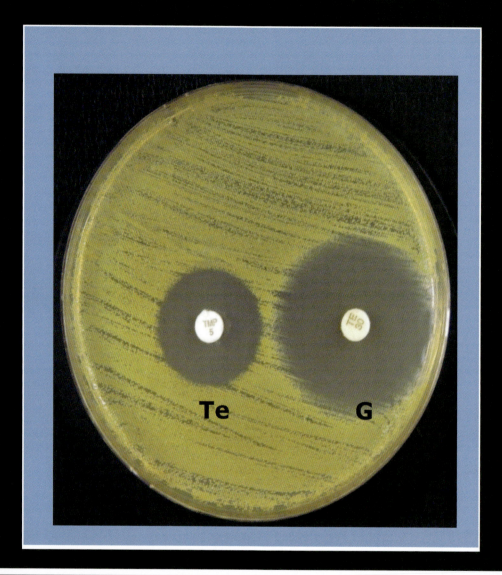

Staphylococcus aureus
Gram-positive
Tetracycline (Te) and
Sulfisuxizole (G)

Chemotherapeutic Agents Synergistic Effect

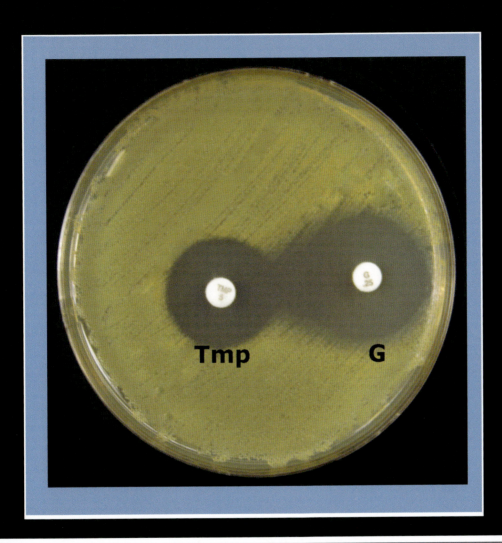

Tmp   G

Staphylococcus aureus
(Gram-positive)

Trimethoprim (Tmp)
and
Sulfisuxizole (G)

Chemotherapeutic Agents: Additive Effect

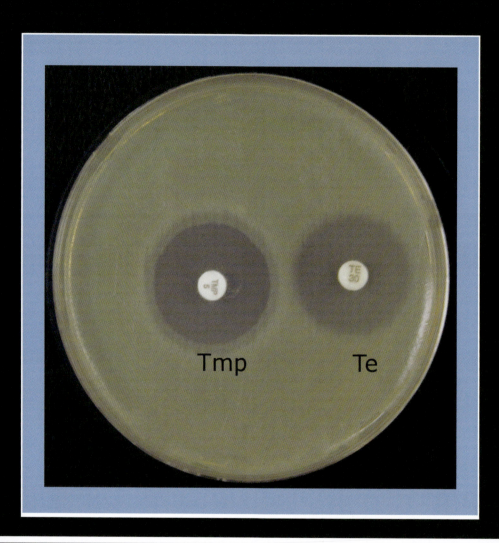

Escherichia coli
(Gram-negative)
Trimethoprim (Tmp)
and
Tetracycline (Te)

Chemotherapeutic Agents
Synergistic Effect

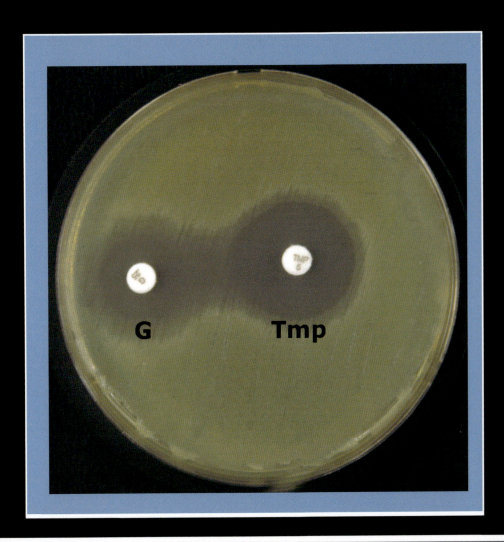

Escherichia coli
(Gram-negative)
Trimethoprim (Tmp)
and
Sulfisuxizole (G)

Chemical Agents of Control:
Antiseptics

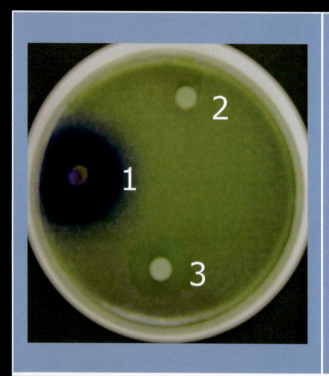
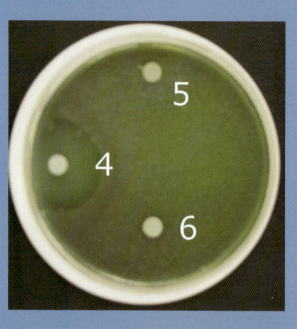

Pseudomonas aeruginosa
(Gram negative)
After 48 hours incubation on Muller-Hinton agar

1= Crystal violet: no zone of inhibition
2= Listerine: no zone of inhibition
3= Tincture of iodine: moderate zone of inhibition
4= Hydrogen Peroxide: Large zone of inhibition
5= 70% isopropyl alcohol: small zone of inhibition
6= Lysol: small zone of inhibition

Chemical Agents of Control:
Antiseptics

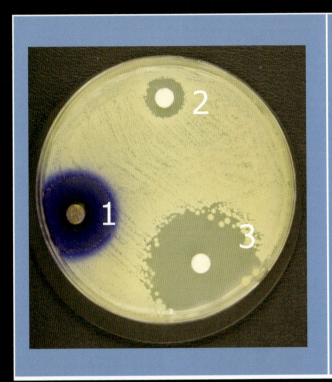 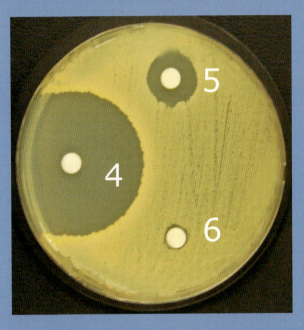

Staphylococcus aureus
(Gram positive)
After 48 hours incubation on Muller-Hinton agar

1= Crystal violet: moderate zone of inhibition
2= Listerine: small zone of inhibition
3= Tincture of iodine: large zone of inhibition
4= Hydrogen Peroxide: Large zone of inhibition
5= 70% isopropyl alcohol: small zone of inhibition
6= Lysol: small zone of inhibition

Chemical Agents of Control:
Antiseptics

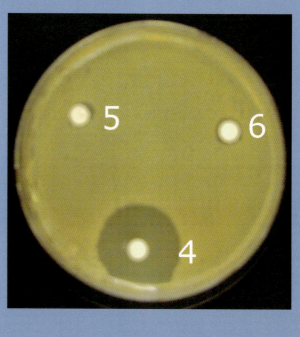

Escherichia coli
(Gram negative)
After 48 hours incubation on Muller-Hinton agar

1= Crystal violet: no zone of inhibition
2= Listerine: small zone of inhibition
3= Tincture of iodine: large zone of inhibition
4= Hydrogen Peroxide: Large zone of inhibition
5= 70% isopropyl alcohol: small zone of inhibition
6= Lysol: small zone of inhibition

Genetics
Microbial Conjugation

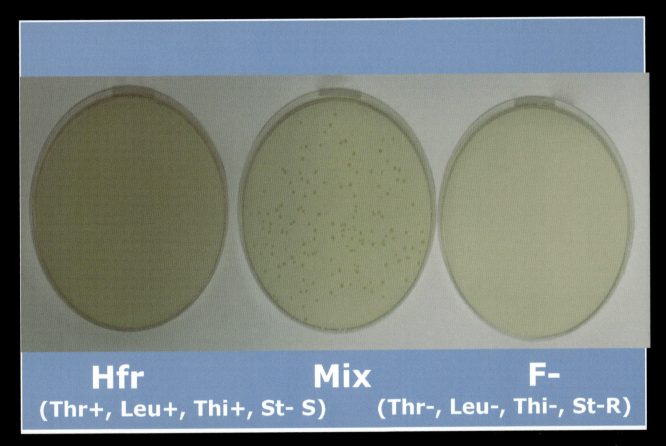

Hfr	Mix	F-
(Thr+, Leu+, Thi+, St- S)		(Thr-, Leu-, Thi-, St-R)

Cultures grown on Minimal Agar with Thiamine and Streptomycin

Hfr = Although the bacterium is Thr+, Leu+, and Thi+ but it is sensitive to streptomycin.

Mix = Growth is positive. This indicates that the F- culture received Thr+ and leu + genes from Hfr bacterium

F- = Although the bacterium is resistant to Streptomycin it is Thr-, Leu-, and Thi-.

IMViC Test
Indol

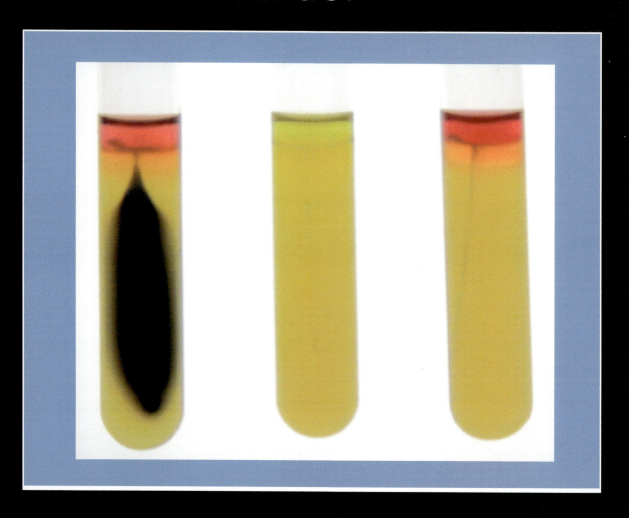

P. vulgaris   E. aerogenes   E. coli

Tryptophane   Tryptophanase   Indol
(substrate) $\longrightarrow$ (End-product)

(Enzyme)

Reagent: Kovacs's reagent

IMViC Test
Simmons Citrate Agar

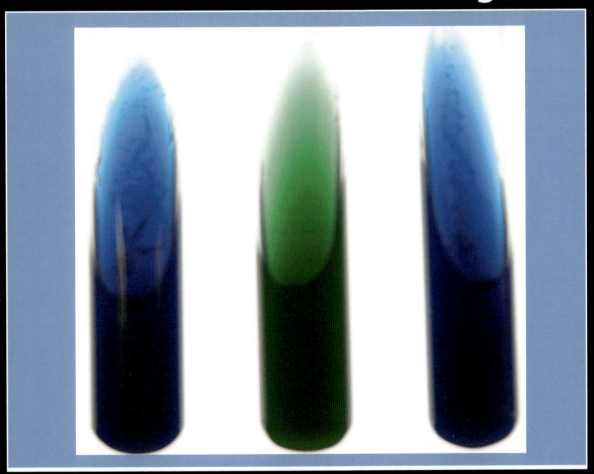

P. vulgaris   E. coli   E. aerogenes

Citrate   citrase   Sodium bicarbonate
(substrate) ⟹ (end-product)
(enzyme)
Reagent: Brom-thymol blue added to the medium

•Green color indicates a negative citrate test
•Blue color indicates a positive citrate test

IMViC Test
MR   VP

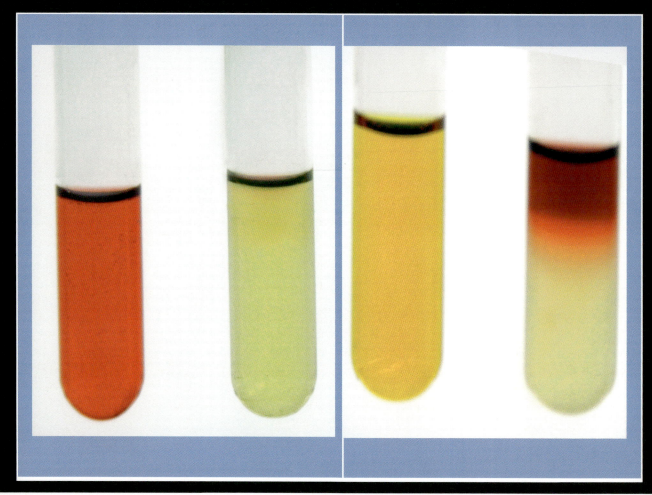

P. vulgaris   E . aerogenes   P. vulgaris   E .aerogenes
E. coli   E. coli

IMViC Test

Methyl Red test

Glucose   ⟶   Acidic end products

(Substrate)   (several enzymes)   (variety of acidic end products)

Reagent/pH idicator:
Methyl red

Voges-Prauskauer test

Glucose   ⟶   Non-acidic end products

(Substrate)   (several enzymes)   (Neutral, acetoin, etc)

Reagent/pH idicator:

1. potassium Hydroxide (KOH) and
2. 2. Alpha-naphthol

Carbohydrate Fermentation
E. coli

Sucrose	Dextrose	Lactose

1. Red broth indicates an alkaline reaction or no carbohydrate fermentation.
2. Yellow broth indicates presence of acid in the tube and utilization of carbohydrate by the bacterium
3. Bubble in the Durham tube indicates the presence of gas.

Carbohydrate Fermentation
Pseudomonas aeruginosa

Sucrose	Dextrose	Lactose

1. Red broth indicates an alkaline reaction or no carbohydrate fermentation.
2. Yellow broth indicates presence of acid in the tube and utilization of carbohydrate by the bacterium
3. Bubble in the Durham tube indicates the presence of gas.

Carbohydrate Fermentation
Staphylococcus aureus

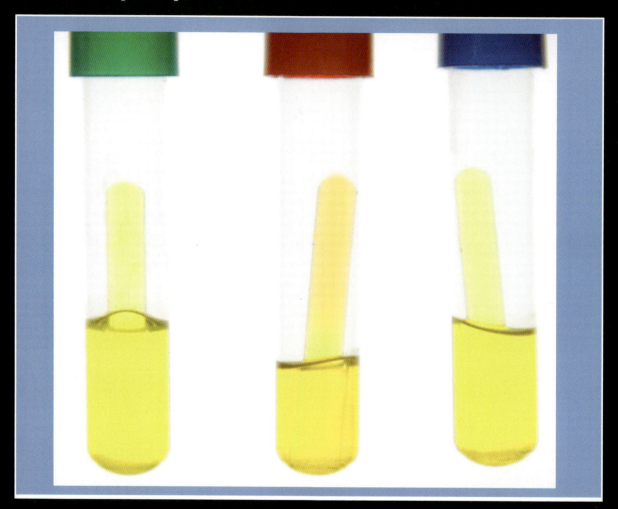

Sucrose   Dextrose   Lactose

1. Red broth indicates an alkaline reaction or no carbohydrate fermentation.
2. Yellow broth indicates presence of acid in the tube and utilization of carbohydrate by the bacterium
3. Bubble in the Durham tube indicates the presence of gas.

Triple Sugar Iron (TSI)

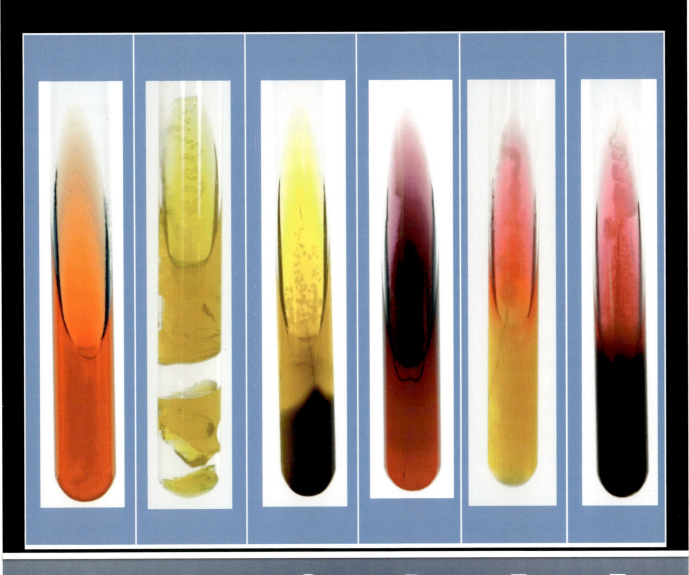

A   B   C   D   E   F

Triple Sugar Iron (TSI)

A = Control

B = E. coli: Glucose, lactose and/or sucrose fermentation . No H2S gas.

C = P. vulgaris: Glucose, lactose and/or sucrose fermentation . Hydrogen sulfide positive

D = P. aeruginosa: No fermentation No H2S gas.

E = S. dysentariae: Glucose fermentation only . No H2S gas.

F = S. typhimurium: Glucose fermentation only. H2S gas positive.

- Yellow color= acidic reaction
- Red =alkaline reaction
- Yellow butt red slant = glucose fermentation only
- Yellow butt and yellow slant = Glucose, lactose and/or sucrose fermentation
- Red butt and red slant = no carbohydrate fermentation
Black butt = hydrogen sulfide gas

Extracellular Enzymes
Gelatin Hydrolysis

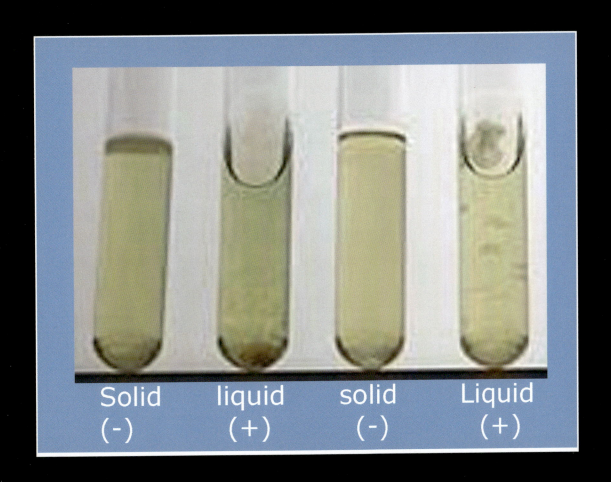

Solid (-)   liquid (+)   solid (-)   Liquid (+)

E. Coli   P. aeruginosa   S. aureus   B. cereus

Gelatin   Gelatinase   **Amino acids**
(substrate)   ⟶   (end-product)
                 (enzyme)

Reagent: none

Extracellular Enzymes Starch Hydrolysis

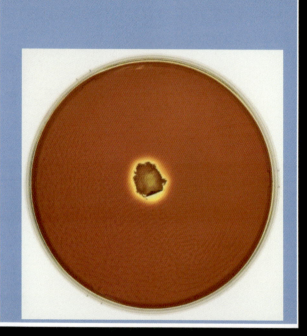

S. aureus
(E. coli   and
P. aeruginosa
also show same
negative reaction)

B. Cereus
(positive)

Starch   **Amylase**   **Maltose**

(substrate)   ⟶   **(end-product)**

(enzyme)

Reagent: Gram's iodine

Extracellular Enzymes
Casein Hydrolysis

P. aeruginosa
(positive: clear halo
around the growth)

B. Cereus
(positive: clear halo
around the growth)

Caseinase

Casein   ➜   **Amino acids**

(substrate)   (enzyme)   (end-product)

Reagent: none

Extracellular Enzymes
Casein Hydrolysis

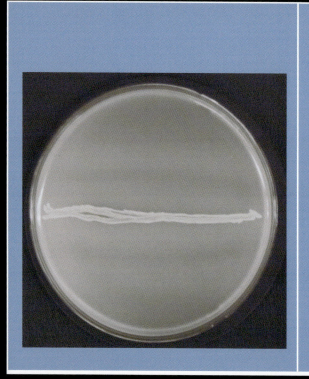 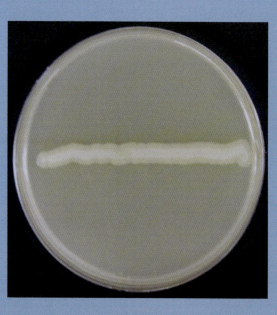

S. aureus
(positive, clear halo
around the growth)

E. Coli
(negative, no clear halo
around the growth)

Caseinase

Casein → **Amino acids**

(substrate)   (enzyme)   (end-product)

Reagent: none

Nitrate Reduction

E .coli   A. faecalis   P. aeruginosa

•3-4 drops of each Solution-A (Sulfanilic acid) and solution-B (alpha-Naphthylamine) has been added to all tubes.

• minute quantity of zinc powder was added to A. faecalis and P. aeruginosa tubes)

Nitrate Reduction

E. coli: turned red after adding solutions A and B indication of a positive nitrate reduction and the presence of nitrite as end product.

A. Faecalis: turned red after the addition of zinc powder indicating a negative nitrate reduction.

P. aeruginosa: remained clear after the addition of all three reagents. This also indicates a positive nitrate reduction but ammonia and molecular nitrogen as it's end products.

Normal Flora of The Skin
Mannitol Salt Agar

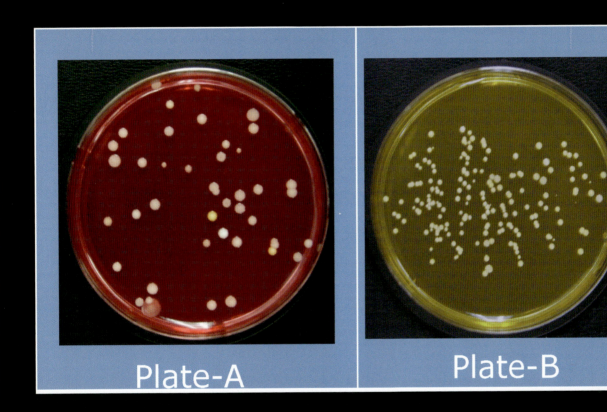

Plate-A

Plate-B

Plate-A: S. epidermidis Color of the Mannitol salt Agar plate remains pink because of lack of mannitol fermentation .

Plate-B: S. aureus Color of the Mannitol salt Agar plate changed from pink to yellow because of mannitol fermentation .

Normal Flora of The Skin
Blood Agar

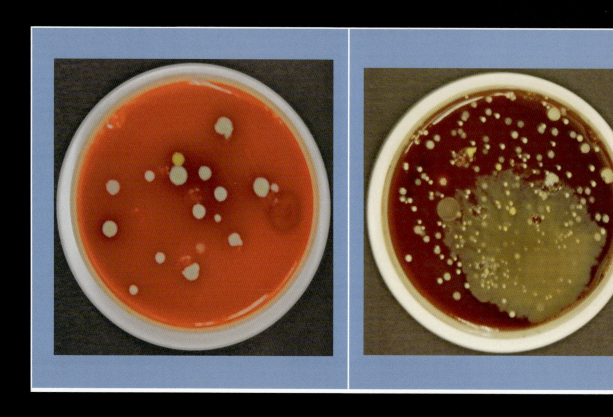

Normal Flora of The Skin

Sabouraud's agar

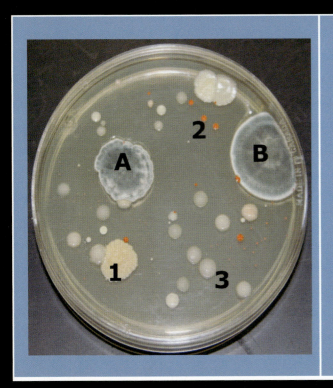 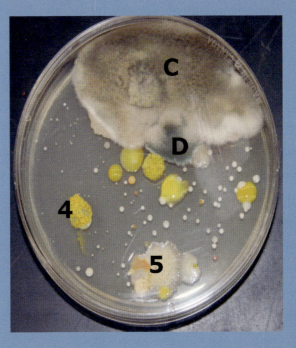

A,B,C,D = Mold (filamentous/multicellular)

1, 2, 3, 4, 5 = Yeast (non-filamentous)

Catalase Test

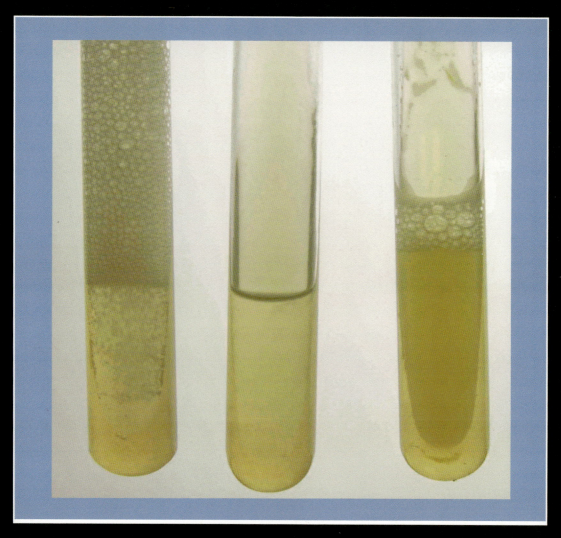

S. aureus   S. lactis   M. luteus

S. aureus: foaming after adding hydrogen peroxide reagent indicates presence of the enzyme catalase

S. lactis: no foaming after the addition of the reagent indicates absence of catalase.

M. luteus: foaming after adding hydrogen peroxide reagent indicates presence of the enzyme catalase

Qualitative Analysis of Milk

Methylene Blue Reduction Test

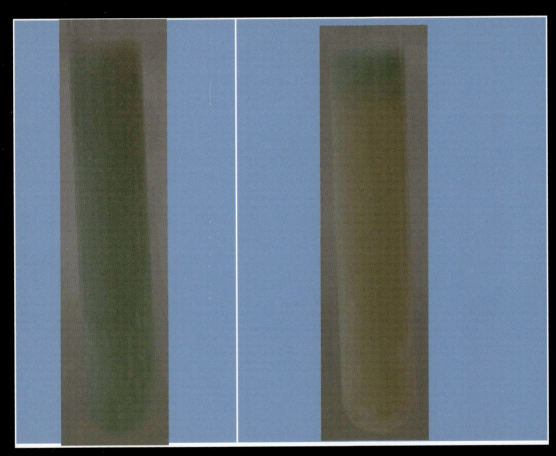

Pasteurized Milk   Raw Milk

Pasteurized milk remained blue after 8 hours of incubation at 37 degrees centigrade. It is classified as "Good quality" milk.

Raw milk turned with in less than 30 minutes of incubation at 37 degrees centigrade. It is classified as "Poor quality" milk.

Index

A

Acid-fast bacterium	7
Acid-fast stain	7
Additive Effect of antibiotics	52
Alcaligenes faecalis	71
Alpha-hemolysis	20
alpha-Naphthylamine	71
Amino acids	67, 69
Ammonia	72
Amoeba proteous	31
Amylase	68
Antibiotics	46
Antiseptics	54
Ascaris lumbricoides	45
Asexual spores	30
Aspergillus niger	27

B

Bacillus	4
Bacillus cereus	4, 6, 11, 25, 68, 69
Bacterial conjugation	57
Beta-hemolysis	20
Blood agar	20, 74
Blood fluke, female	40
Blood fluke, male	39
Brom-thymol blue	59
Broth culture	2
Bud	26

C

Capsule	9
Capsule stain	9
Carbohydrate fermentation	62, 64
Casein	69, 70
Casein hydrolysis	69, 70
Caseinase	69, 70
Catalase test	76
Cestoda	35, 36, 37
Chemotherapeutic agent	46
Cilliophora	34
Citrase	59
Citrate	59
Clear zone	20
Clostridium botulinum	8
Clostridium sporogenes	22, 23
Coccus	4
Colonies	10
Conidiospores/conidia	30
Counter stain	5
Cryptococcus neoformans	26
Crystal violet	4, 5, 54-56
Cultural characteristics	10
Cyst	32

D

Decolorization	5
Deep tube	2
Dextrose	62-64
Diplobacilli	4
Dog tapeworm	35, 36
Durham tube	62

E

Early schizont	33
Endospore	8
Endospore stain	8
Entamoeba histolytica	32
Enterobacter aerogenes	18, 58, 59, 60
Enterobius vermicularis, female	42
Enterobius vermicularis, male	41
Enterococcus faecalis	20, 21
Enumeration of bacteria	14
Eosin-Methylene Blue Agar	19
Escherichia coli	3, 4, 10, 18, 19, 21, 22, 23, 48, 52, 53 , 56, 58, 60, 62-66, 70, 71
Extracellular enzymes	67-70

F

F-culture	57
Fasciola hepatica	38
Fungi	26, 27, 28, 29, 30

G

Gamma-hemolysis	20
GasPak jar	23
Gelatin	67
Gelatin hydrolysis	67
Gelatinase	67
Genital pore	36, 37
Giardia lamblia	32
Glycoprotein	9
Gonads	36
Gram negative	5, 18, 19, 21, 24, 47
Gram positive	5, 18, 19, 21, 24, 46, 50
Gram stain	5
Gram variability	5
Gram's iodine	68
Green-metallic sheen	19
Greenish zone	20
Gynaecophoric canal	39

H

Halophilic bacterium	17
Hemoglobin	20
Hfr culture	57
Hooks	37
Hookworm, female	44

Hookworm, male	43
Human blood	1
Hydrogen Peroxide	54-56, 76
Hydrogen sulfide gas	66

I

IMViC test	58-60
Indol test	58
Isolated colonies	10
Isopropyl alcohol	4-56

K

Klebsiella pneumoniae	9
Kovac's reagent	57

L

Lactose	18, 19, 62-64
Late schizont	33
Lipid	5
Listerine	54-56
Liver fluke	38
Lysol	54-56

M

Magnification	1
Maltose	68
Mannitol	17
Mannitol fermentation	17, 73
Mannitol salt agar	17, 73
Mastigophora	32, 34
McConkey agar	18
Melachite green	8
Merozoite	33
Methylene blue	4
Methylene blue reduction test	77
Micrococcus luteus	3, 6, 12, 22, 23, 76
Minimal agar	57
Mold	26, 27, 28, 29, 30, 75
Molecular nitrogen	72
Mycobacterium smegmatis	7, 49

N

Necator americanus, female	44
Necator americanus, male	43
Neck	37
Negative stain	6
Nemathelminthes	40-45
Nematota	40-45
Nigrosin	6
Nitrate reduction test	71
Nitrite	72
Non-acid-fast	7
Nutrient agar plate	10
Nutrient agar slant	10
Nutrient broth	10

O

Objective lens	1
Ocular lens	1
Oral sucker	38, 39, 40
Ovary	36, 40

P

Parasitic helminthes	35-45
Parasitic protozoa	32, 33, 34
Pasteurized milk	77
Penicillin	46-49
Penicillium notatum	30
Peptidoglycan	5
Phenylethyl alcohol medium	20
Pinworm, female	42
Pinworm, male	41
Plasmodium vivax	33
Platyhelminthes	34, 35, 36
Pleomorphic	7
Polysaccharide	9
Primary Stain	5
Proglottid	36, 37
Proteins	9
Proteus vulgaris	58, 60, 66
Protozoa	31
Pseudomonas aeruginosa	13, 47, 54, 63, 66, 69, 71

Q

Qubec Colony Counter	16

R

Raw milk	77
Rhizopus nigricans	1, 28, 29
Rostelum	37

S

Sabouraud's agar	27, 30, 75
Safranin	4, 5
Salmonella typhimurium	19, 66
Sarcodina	32
Schistosoma japonicum, female	40
Schistosoma japonicum, male	39
Scolex	37
Selective and differential media	17
Serial dilution method	14, 15, 16
Serratia marcescens	2
Sexual spore	29
Shigella dysentariae	66
Signet ring	33
Simmon citrate agar	59
Simple stain	4
Slant culture	2
Sodium bicarbonate	59
Sodium thioglycollate broth	22
Sporozoa	34

Staphylococcus aureus   17, 21, 24, 46, 50,
                                    51, 55, 64, 70, 76

Staphylococcus epidermidis   4, 17
Starch hydrolysis   68
Streak Plate Method   3
Streptococcus lactis   76
Streptococcus mitis   20
Streptococcus pyogenes   20
Streptomycin   46-49
Suckers   37
Sucrose   62-64
Sulfanilic acid   71
Sulfisoxazole   46-51
Synergistic Effect   53

T

Taenia pisiformis   35, 36
Testes   36, 37, 39
Tetracycline   46-49
Tetrads   6, 7
Thiamine   57
Tincture of iodine   54-56
Too Few to Count (TFTC)   15
Too Numerous to Count (TNTC)   14, 15
Toxoplasma gondii   34
Trematoda   39, 40
Trimethoprim   51, 52, 53
Tripanosoma gamgiense   1, 34

Triple sugar iron test   65, 66
Trophozoite   32, 33, 34
Tryptophanase   58
Tryptophane   58

U

Ultra violet radiation   24
Uterus   36, 37, 40

V

Vagina   36
Vancomycin   46-49
Vas deferens   36
Vegetative cells   8
Ventral sucker   38, 39, 40
Virulent   9
Vitelline gland   36, 37, 40
Voges-Prauskauer test   60

Y

Yeast   26, 75

Z

Zinc power   71
Zone of inhibition   54-56
Zygospore   29